PHYSICS AT A GLANCE

By

Ogechukwu NNADIH

First published by
Obafemi Awolowo University Press
Ile-Ife, Nigeria.

ISBN: 978-136-393-2

TABLE OF CONTENT

DEDICATION

This book is dedicated to God Almighty who inspired me and to all who will find this complimentary guide to Physics, useful in their quest to understand Physics.

ACKNOWLEDGEMENTS

Writing this book has been a great challenge and also a learning experience from the time the idea was conceived till date. It all started like a dream; to encourage students to study Physics, understand its concepts and also relate the physical phenomena in it (physics) to our environment.

It would have been impossible to complete this work without the help, suggestions and support I got from so many scholars; most especially the likes of Professor P.N Okeke (Department of Physics and Astronomy, University of Nigeria Nuskka), Professor Ama Nduka (Department of Physics, Federal University of Technology Owerri), Mr. E.N. Gajere (The Director, National Centre for Remote Sensing, Jos), Dr. A. C Ugwoke and Mr. Onyia (Department of Physics, Enugu State University of Science and Technology) Dr. B.I Okere (Centre for Basic Space Science, Nuskka), Professor E. Obiajunwa (Professor of Physics, Centre for Energy Research and Development, Obafemi Awolowo University, Ile-Ife, Osun State) and Sir (Engr) S.U.B Ezekpo (Department of Electrical Electronic Engineering, Obafemi Awolowo University, Ile-Ife, Osun State), who painstakingly re-edited the work. May God reward and bless their families abundantly.

To all my lecturers in the Department of physics, Enugu State University of Science and Technology, members of the Nigerian Institute of Physics (NIP) and Institute of Physics (IOP) community for their numerous support and encouragement, I say a big thank you.

Finally I also wish to express my profound gratitude to God who made all these efforts successful; to my parents, and siblings, I am really proud of you; friends, critics. May God reward everyone according to their deeds

PREFACE

A study of Physics and its application is basic and vital to all students irrespective of their educational goal. Most often, students do ask "How do I study and understand Physics? How do I cope with its formulae and proofs? How do I prepare for its examination? How do I make "A's", manage "B's" and reject "C's" in physics?"

Physics at a glance is intended to give an ordinary person a brief overview, the importance and wonder of Physics around us and in the Universe at large. Sometimes people think we need the mind of *Isaac Newton, Albert Einstein, James Clark Maxwell,* or other of the great scientists in order to understand Physics, its formulae and applications, so they give up on it entirely.

First and foremost, we should maintain a positive attitude towards the subject, keeping in mind that Physics is the most fundamental of all natural sciences where all other science courses that follow use the same physical principles as in Physics.

This book is also intended for students in secondary schools, Physics Departments in our higher Institutions, Engineering departments in our universities and polytechnics and in other related sciences. Most especially, the target of this book is to create a positive mental attitude towards the subject for those who would not like to study the course, but applying its laws and principles in one thing or the other, *that Physics is a challenging and rewarding subject to study*. Its study instructs a person in the art of critical reasoning, how to pose questions and solve problems.

Physics is at the heart of almost every aspect of modern life. It provides training for a vast range of careers, where it is either employed directly or its skills developed and applied in innovative ways in other fields. Hence, it is important we study Physics, understand its principles and be able to apply them in our daily lives.

FORWARD

Physics as a natural science that studies matter and its motion in space and time, gives man an understanding of the world he lives in and how best to exploit it for his benefits.

The author of this book "**Physics at a glance**" has put in our hands a book that gives a brief but clear overview of Physics without delving into the complex formulae and mathematics associated with the subject. Thus the book is designed to give the ordinary person (non physicist) a gist of the subject matter in a readable, easy to understand format. The book is also presented in such a way as to arouse the interest of the reader to dig deeper into the subject.

The chapter on "**Suggested ways to study and understand Physics**" is a very useful guide to students of Physics on how to get the best out of the subject. The chapter on "**Applications of the knowledge of Physics in our daily life**" shows clearly how we interact with Physics and its off-shoots in our everyday life.

I highly recommend this book first, to the general public for a better understanding of the world we live in and to Physics students in particular for the tips in understanding Physics contained therein.

Professor E. I. Obiajunwa (Professor of Physics),
Centre for Energy Research and Development,
Obafemi Awolowo University,
Ile-Ife,
Nigeria.

CHAPTER ONE

AN OVERVIEW OF PHYSICS

1.1 *PHYSICS AS A SUBJECT*

The word Physics came from a Greek word meaning *"Knowledge of Nature"*. It attempts to describe the fundamental nature of the universe and how it works.

According to Dull .C. (1960), Physics is as old as ancient Greece and yet as new as tomorrow's newspaper. It is the study of the laws of nature that govern the behavior of the Universe.

Murphy .T. (1982) in his book "Physics Principle and Problem" defined it as the branch of science that involves the study of physical phenomena in order to establish patterns. Hence, it is the most basic and fundamental of all sciences and studies the nature of matter, energy and their relationship, which involves theories and experiments (Experimental Science).

The experiments and theories in physics are explained by a small number of relations, which are often expressed using mathematical symbols, thus, *the language of physics* is mathematics.

The physical phenomena in our world of physics are a part of the following areas in Physics;

1. MECHANICS
2. STATISTICAL MECHANICS
   o THERMODYNAMICS
3. ELECTROMAGNETISM
4. RELATIVITY AND ASTROPHYSICS
5. QUANTUM MECHANICS
6. WAVE AND OPTICS

1.11   MECHANICS

Mechanics is a branch of Physics that is concerned with motion and forces that cause it. It equally includes the study of mechanical properties of matter such as density, elasticity and viscosity. Mechanics may be divided into *Static* and *Dynamics*. Static deals with bodies at rest while dynamics deals with bodies in motion and sometimes is further sub-divided into *Kinematics* (which describes motion without regard to its cause) and *Kinetics* (which explains the changes in motion as a result of the force on it).

The principles in Mechanics are based on the laws of Newton, which state as follows:

I.   A body at rest will continue to be at rest, while a body in motion will be in motion unless acted upon by an external force.

II.   If a net force acts on a body, it accelerates; the direction of acceleration is the same as direction of the applied force.

III.   Action and reaction are equal and opposite.

Mechanics can generally be divided into *classical mechanics* and *quantum mechanics*. Classical in the sense that it considers the motion of large objects while quantum mechanics considers the motion of tiny particles such as an atom.

1.12   STATISTICAL MECHANICS

This is a branch of Physics that combines the principles and procedures of Statistics with the laws of both classical mechanics and quantum mechanics. It considers the average behavior of a large number of particles rather than the behavior of any individual particle.

Statistical mechanics recognizes three broad types of system; those that obey Maxwell-Boltzmann Statistics, those that obey Bose-Einstein Statistics and those that obey Fermi-Dirac Statistics.

Maxwell Boltzmann Statistics apply to systems of Bosons (particles that have integral values of quantum mechanical properties called spin). Fermi-Dirac Statistics apply to system of fermions (particles that have half integral values of spin).

Statistical mechanics provides a framework for relating the microscopic properties of individual atoms and molecules to microscopic properties of materials that can be observed in everyday life, therefore explaining thermodynamics as a result of statistics and mechanics (classical and quantum) at the microscopic level. It is also known as **Statistical Thermodynamics**.

Thermodynamics on its own is a branch of Physics that studies energy transformation involving heat and mechanical works. It forms an indispensable part of the foundation of Physics, Chemistry and Life Sciences. It is applied in such places as car engines, refrigerators, air conditioners, biochemical processes and structures of the stars.

In car engine, heat is generated by the chemical reaction of oxygen and vaporized gasoline in the engines cylinders. The heated gas pushes on the piston within the cylinders doing mechanical work that is used to power the car. Refrigerators on their own side absorb heat from a cold place (inside the refrigerator box) and give it out to a warm place (in the room where the refrigerator is located), thus low ventilated apartments with refrigerator are usually warm.

The Air Conditioner operates on the same principles as in the refrigerator, but in this case, the room or entire building becomes the refrigerating boxes. Its mode of operation is beyond the target of this book.

All these thermodynamics processes are guided by certain rules or laws called the laws of thermodynamics, which are as follows:

A. **Zeroth Law**

"If two objects A and B are in thermal equilibrium with a third object C, all of them are in equilibrium with one another".

Its importance was recognized only after the first, second and third laws were named, and since it is fundamental to all of them, it is now called the **Zeroth law**.

B. **First Law**

"It is impossible to construct a continuously operating machine system that does work without obtaining energy from an external source".

C. **Second Law**

"It is impossible to construct a continuously operating system, that obtains energy from an external source (reservoir) and converts it completely into mechanical work"

D. **Third Law**

"It is not possible to lower the temperature of any system to absolute zero in a finite number of steps".

1.13 ELECTROMAGNETISM

Electromagnetism is a branch of science that is dealing with observations and laws relating electricity to magnetism. It is based on the fundamental observations that a moving electric charge produces magnetic field, and that a charge moving in magnetic field will experience a force. The magnetic field produced by a current is related to the current, the shape of the conductor and the magnetic properties of the medium around it by Ampere's law.

There are still other laws and theories propounded by other scientists that explain electromagnetism such as Coulombs law by **Charles Augustine de Coulomb** (1736-1806), which stated that

> *"The force of attraction between two point charges is directly proportional to the product of their magnitude and inversely proportional to the square of the distance between them"*.

Gauss's Law by Carl Fridrich Gauss (1777-1855) stated that

> *"The total flux* (collection of field lines) *passing an enclosed surface is directly proportional to the total electric charge inside the surface"*.

It is a fundamental statement for the relationship between electric charges and electric field. Its consequence is that static charges (charges at rest) on a conductor are formed on the conductors surface not on its interior. Thus, when one touches a charged metal, he may be easily electrocuted.

1.14 RELATIVITY AND ASTROPHYSICS

Until the end of Nineteenth Century, physicists believed that all physical phenomena ranging from the motion of atoms to the celestial bodies were governed by one set of laws; The laws of motion were formulated by Newton in his book known as the **Principia**. Newton's theory implies that these laws are valid for systems that move at constant speed relative to each other. The law, which states the same for such system, is known as **Principle of Relativity.**

The Theory of Relativity is special (Special Theory of Relativity) in the sense that it treats the invariance of the laws of nature principally for uniformly moving frames, i.e. Frames that are not accelerated.

5

The most remarkable aspect of the theory is Einstein's Law of Mass-Energy Equivalence, i.e. mass and energy are one and the same because when something gains energy, it gains mass as well and when it loses energy, it loses mass.

The Theory of Relativity is based on two simple postulates viz:

I.   *"The laws of physics are the same in all inertial frame of reference"*.

II.   *"The speed of light in vacuum is the same in all inertial frame of reference and is independent of the motion of the source"*.

Einstein's conviction that the laws of nature should be expressed in the same frame of reference was the primary motivation that led to the General Theory of Relativity – a theory which states that the effect of gravity and acceleration cannot be separated from one another. An interesting consequence of this theory (Special Theory of Relativity) is that space and time are no longer viewed as separate independent entities but rather, are seen to form a four-dimensional continuum called **Space-Time**. For instance, we describe a box by its length, width and height. But the three-dimension does not give a complete picture of the box, there is a fourth dimension-time, the box was not a box at a given length, width and height. It began as a box only at a certain point in time, on the day it was made and will cease to be a box on the day it might be destroyed.

We cannot speak meaningfully about space without implying time, things exist in space-time, each object, person etc exists in what physicists called **Space-Time continuum**.

1.15   QUANTUM MECHANICS

Quantum Mechanics is a branch of Physics that describes the motion of small objects like the atom and subatomic particles. It is the theoretical framework within which it has been found possible to describe and predict the behavior of a vast range of physical systems from elementary particles through nuclear atom and radiation to molecules and solids. Its fundamental concepts are the Schrodinger Equation as Maxwell's equations are to Electromagnetism.

Quantum mechanics has had great success in explaining and predicting new behavior that is universally accepted today; understanding the laws of quantum mechanics has led to advances in many areas of Technology such as development of Computer Memory Chips and has increased our understanding of the phenomena ranging from the Molecular Basics to Biology of the life cycle.

1.16   WAVES AND OPTICS

A wave is the transfer of energy by regular vibration or oscillatory motion. They are classified into *Longitudinal or Compressional waves* such as sound waves, which travel through a material medium by alternatively forcing the molecules of the medium closer together then spreading them apart, and *transverse waves* such as light waves and all other forms of *electromagnetic waves* in which the vibration is at right angles to the transfer of energy.

1.17   ATOMIC AND NUCLEAR PHYSICS

Atomic Physics is a field in Physics that studies atoms as isolated system, comprising electrons and an atomic nucleus. It is primarily concerned with the arrangement of electrons around the nucleus and the processes by which these arrangement changes. Atomic Physics also studies the structure of atom, its dynamics properties including energy states and its interactions with particles and fields.

The true beginning of Atomic Physics is marked by the discovery of **spectral line** and attempts to describe the phenomena (spectral line).

Nuclear Physics on its own is a branch of Physics that is concerned with the nucleus of an atom. It has three aspects such as investigating the fundamental particles (protons and neutrons) and their interactions, classifying and interpreting the properties of nuclei, and providing technological advances.

The term Atomic Physics is often associated with nuclear power and nuclear bombs, due to the synonymous use of atomic and nuclear physics in Standard English. However, physicists distinguish between:

i. Atomic physics – which deals with the atom as a system of electron(s) and nuclei
ii. Nuclear Physics – which considers atomic nuclei alone

Fully detailed explanation and descriptions of these concepts (Mechanics, Statistical Mechanics, Electromagnetism, Relativity and Astrophysics, Quantum Mechanics, Wave and Optics, Atomic and Nuclear Physics) are beyond the scope and target of this book.

Some of the main fields in Physics are:

i. Acoustics (the science of sound)
ii. Atmospheric and Space Physics
iii. Atomic and Molecular Physics
iv. Biophysics
v. Condensed Matter Physics
vi. Cryogenics
vii. Electrodynamics
viii. General Relativity and other gravity theory.
ix. Geophysics
x. Nuclear and Particles Physics
xi. Solid State Physics

1.2 THE IMPORTANCE OF PHYSICS

Physics is important to all other Sciences; Chemistry deals principally with the applications of the laws of Physics to the formation of molecules and different practical ways of transforming certain molecules to another. Biologists depend heavily on Physics and Chemistry to explain the processes occurring in living bodies.

Engineering is nothing but the application of the principles in Physics and Chemistry, for instance, Physics gave birth to Electronics, which has technologically changed the world. Hence, Physics is not only contributing to the knowledge about nature but also to the social progress of Mankind.

1.3   SUB-DIVISIONS IN PHYSICS

Physics can be divided into classical Physics, where all the laws of Newton's and theories of other scientists like Kepler, Galileo and so on are valid, and Quantum physics which was established by Werner Heisenberg, Max Plank, Paul Dirac and others.

Classical Physics was the first branch of Physics to be discovered and also the foundation upon which other branches of Physics were built. It studies motion of gasses, liquids and solids, but more commonly, it is taken to refer only to solids. In the restricted reference to solids, Classical Physics is sub divided into:
  I.   Static
  II.   Kinematics
  III. Dynamics.

STATIC considers the action of forces that produce equilibrium or rest.

KINEMATICS deals with the description of motion without concern of the causes of the motion.

DYNAMICS involves the study of objects under the action of the force upon them.

Classical Physics has many applications in other Sciences such as
  i.   Astronomy
  ii.   Chemistry

iii. Geology
iv. Engineering

Quantum Physics on its own is the study of matter and radiation at the atomic level. The word Quantum refers to discrete unit (a theory which states that energy is in discrete form, not continuous). The discovery that waves have discrete energy packets (call quanta) that behave in a manner similar to particles led to the branch of Physics that deals with atomic and subatomic systems which today are called *Quantum Mechanics*. It is the underlying mathematical framework for many fields of Physics and Chemistry, Quantum Chemistry, Atomic Physics and Chemistry, Particle Physics and so on. The foundations of Quantum Mechanics were established during the first half of the twentieth century by Werner Heisenberg, Erwin Schrodinger, Max Born, Niels Bohr, Louis de Broglie and others.

Historically, Quantum Physics was developed using intuition and analogy with classical physics. Even so, there are basic differences in philosophy between them. For instance, for a system obeying classical mechanics (although this is approximation), if all the positions and velocities of the particles are known at some instant, then the state of the system at some later time is completely determined by Newton's Laws. This is not the case for a system obeying quantum mechanics. In quantum physics, the position and momentum of a quantum particle can not be known exactly at the same time.

1.4 **DIFFERENCES BETWEEN CLASSICAL AND QUANTUM PHYSICS**

Quantum Physics is different from classical physics in the sense that most important concepts explained but it cannot be explained by classical physics. These include;

I. DISCRETENESS OF ENERGY
II. WAVE PARTICLE – DUALITY OF LIGHT AND MATTER

10

III.   QUANTUM TUNNELING
IV.   HEISENBERG UNCERTIANITY PRINCIPLE

1.41 **DISCRETENESS OF ENERGY**

This theory explains that energy exists in a small unit of definite amount. For instance, in an atom, when electrons in the quantum state orbit round the nucleus, each orbit is of a defined energy level. Thus, when an electron in a higher energy state jumps down to a lower one, as principle of conservation of energy is concerned, the electron emits packets of energy (proton), which correspond to the exact energy difference between the higher and lower levels.

This explanation defiles the classical theory, which predicted that the electrons would crash at the center of the nucleus.

1.42 **WAVE PARTICLE – DUALITY OF LIGHT AND MATTER**

In 1690, **Christian Huygens** theorized that light was composed of waves. In 1704, **Sir Isaac Newton** explained classically that light is made up of tiny particles; Experiments supported each of these theories, however neither a complete wave theory nor particle theory could explain all the phenomena associated with light, thus, scientists began to think of light as both particles and waves.

In 1923 **Louis de Brogile** stated that radiation as a particle could also exhibit wave-like properties. As particles it has position, momentum, kinetic energy, mass and so on. While as wave, it has wavelength, frequency, amplitude and so on as its properties.

These two behaviors are complementary to each other; thus, he reconciled them by stating that an electromagnetic radiation has dual nature. **Particle and wave nature.**

1.43   QUANTUM TUNNELING

Classical Physics believes that nothing can pass through a barrier without damaging it. That is, it is impossible to throw a particle against a wall and it penetrates through it. Quantum Physics however, states that there is a small probability that this particle could go right through the wall (without damaging it) and continues its flight on the other side, but this probability is so small that one can throw the particle a billion times and never see it pass through the wall, but with some particles as tiny as an electron, this probability is significant.

1.44   HEISENBERG UNCERTAINTY PRINCIPLE

Classically, people are familiar with measuring things around them; someone pulls out a tape measure and determines the length of an observable (object). At the atomic scale of quantum mechanics, measurements become a delicate one because if an observation is measured, the act of measurement may disturb the system and change the value of some other observable quantities.

Wener Heisenberg was the first to realize that certain pairs of measurements have an intrinsic uncertainty associated with them; for instance, if one has a good idea of where some thing is located, to a certain degree, one may have a poor idea of how fast it is moving or in what direction, thus he stated that:
"It is impossible to measure the position and momentum of a particle at the same time".

Heisenberg's Uncertainty Principle completely flies in the face of classical physics, though the foundation of science is the ability to measure things accurately, but Quantum Mechanics is saying that it is impossible to get these measurements accurate if the variables are arbitrary.

CHAPTER TWO

SUGGESTED WAYS TO STUDY AND UNDERSTAND PHYSICS

Studying is different from reading, you can read magazines, newspapers, journals and novels while lying down on your bed, in a bus, anywhere or at anytime, but you can't read physics but have to study it to apply it to your day to day activities.

In this book, you have seen a little of what Physics is all about. For you to study Physics, you need to know the meaning of the keywords, which are either written in bold-faced letters or in italics when first introduced or defined. You should write out the definition of each term, just as you write the vocabularies when reading a new language; doing so will help you learn their precise meanings.

Studying Physics requires more than just learning new terms, definitions and stating the concepts and laws. It implies applying these concepts to real and hypothetical situations. You should always study with pen or pencil (what I mean is different from understanding or highlighting parts or pages of the text) but outlining the ideas as they are presented and working out the algebraic steps for yourself.

When studying and working for yourself (solving problems in Physics), you should always help yourself by arranging your work in an orderly manner, don't scatter equations around the pages of your book. Using a systematic and orderly procedure will help as you are trying to tell your mentor (teacher) about your solution.

Always include the units (units of the physical quantities you are dealing with) when substituting a value for a number in an equation. It is wise to make some comments, literally (using sentences) in your work to indicate what you are doing and why you are doing so. This will make it easier for you when studying or reviewing and will equally help your teacher to understand your work, in case you are writing an examination.

When you study to the end of a section or chapter. Ask yourself questions such as:
   I.   Did I really understand what I read?
   II.   What key ideas and terms have I studied?

Write them down, compare your answers with what you have in the text; solve the problems or exercises at the end of the chapter. Doing so will help you remember what you learnt and will put smiles on your face in the world of Physics.

2.1   PROBLEM SOLVING STRATEGIES IN PHYSICS

For you to achieve your aim in the study of Physics and make maximum use of your time, you must recognize that problem solving is much more than merely substituting numbers to a formula, you should begin by studying the ideas, concepts and relationships first, then attempt the question (problems) that follows to find out if really you understood the passage you read.

When you study a chapter, study as well the text examples; work them out carefully, with pen and paper in a detailed and concise form because practice makes improvement, and to succeed in Physics is by really trying.

Afterwards, follow these steps to know if physics would not be as easy as an "ABC' for you.

   I.   Read the entire problem carefully, this will enable you know what you are being told, don't worry about the question at first, focus more on what information you are being given.
   II.   If possible, draw a diagram of the physical situation because Physics is all about physical things we encounter

everyday. Label the diagram with the information given in the problem, including the units such as (m) for meter, (kg) for kilogram, (s) for seconds, the units of time (t) and so on with the quantities; you may make a list of the known and the unknown quantities involved.

III.   When you are sure of understanding what was given and have labeled your diagram, <u>tackle the question</u>.

IV.   The next step is to <u>find the mathematical relationship between the known and the unknown quantities</u>. In most cases, you need to write the relationship in form of equation(s).

V.   Next, you should <u>solve the equation(s) for the unknown quantities</u>, i.e. rearranging the formula in accordance with the rules of Algebra so that you have the equation(s) with the unknown on the left-hand side of the equality sign and the known quantities and constant (if any) on the right hand side.

VI.   <u>Substitute the numerical values into the equation(s)</u>. Do not substitute just 'bare numbers' (ordinary numbers), but both numerical values and their units. Units are multiplied and divided as if they were algebraic quantities, your answers would then come out with the appropriate units.

Equally, answering these questions will also help you to improve in your world of Physics:

a.   Am I able to use fundamental mathematical concepts in Physics? If not, plan a review course with your teacher.

b.   What has been the easiest area for me? (Study those areas first to build confidence) and which areas in Physics have given me the most trouble? (Spend more time on them).

c.   Do I work in a quiet place where I can maintain my focus? This is because distractions can cause you to miss important points.

d.   Do I understand the material better if I read the book before or after lecture? (You may learn best by skimming the materials, going to lecture and undertaking an in-depth reading after lecture).

2.2   WHY WE SHOULD STUDY PHYSICS

Physics is crucial to understanding the world around us, the world inside us and the world beyond us.

It plays an important role in the scientific quest to understand how human activities affect the atmosphere. Scientists of all disciplines and Engineers make use of the ideas in Physics in the design of cars, electronic sets, computers and so on; thus, it is the foundation of all Engineering and Technology.

Physics is an essential part of science and one of the most fascinating fields of study, its laws are fundamental to all physical sciences and related field such as Medicine, Electronics, Oceanography, Metrology, Material Science and so on. Its importance can be felt in all areas of life. The Clinical Thermometers, X-ray Machines, Camera, Radios and so on, are few of the many inventions and discoveries of men that require knowledge of Physics for their understanding.

Physics education gives a student a solid foundation in basic science. In addition, it develops his or her analytical thinking skills to a higher level and equally equips the student to work in many different and interesting places, on college campuses and so on.

So… Physics is interesting, relevant and can prepare you for that great job in a wide variety of places.

CHAPTER THREE

APPLICATIONS OF KNOWLEDGE OF PHYSICS IN OUR DAILY LIFE

Knowledge of physics is applied virtually in everything we do, in our homes, environment and the universe at large. We find its application in our television sets and fluorescence tubes, in our homes as a result of the action of cathode rays, invented by **Sir William Crooks.** The knowledge of Physics is equally applied in the Cellular Phone as well as the Radio through Electromagnetism, whereby we receive signals in both Amplitude Modulation (AM), frequency modulation (FM) in case of a radio and other Modulation methods such as Pulse Code Modulation (PCM).

Have we ever wondered to ask ourselves why the Power Holding Company (PHC) preferred to sag the high-tension wire, instead of making it taut? Why the sky is apparently seen as light blue in color and not any other color? What causes the thunderclap and how small thunder protects our storey buildings and offices from thunder strike? Or even the mechanism by which we receive calls on our phones, and radio waves for our radios? All these are possible puzzles in life, which can be explained by the world of Physics, as follows:

3.1   THE BLUE COLOUR OF THE SKY

The atmosphere as we know it is filled with dust particles and water moisture. We can notice the particular components of the atmosphere once we see the sun's ray reflecting from a source. We see the dust particles moving to and fro. We can equally notice the moisture component of the atmosphere when we feel the droplets of water on our environment in the early hours of the day. These explain to us that the atmosphere is filled with dust particles and water moisture.

As the sun's rays strike on these components (particles and moisture) they produce refraction of the colors of light (Red, Orange, Yellow, Green, Blue, Indigo and Violent) popularly known as ROYGBIV, with the scatter of shorter wavelength of the suns rays. The colors with the least wavelength (Blue and Violet) produce an illusion of blue sky.

On a normal circumstance, the actual color of sky is seen in the night as black outside the reflection of light from the stars (sun). The blue color we see is an apparent color while the real color is black as we see at night, thus, **the sky is black not blue**.

3.2 THE WORKING PRINCIPLE OF THUNDER PROTECTOR

In Physics, there is what is called **Action of point phenomena** i.e. migration of charges to the region of sharp curvature. The thunder protector as we know has "finger-like" Copper which is usually buried inside the earth, and the earth, as we know, equally has charges (both positive and negative charges).

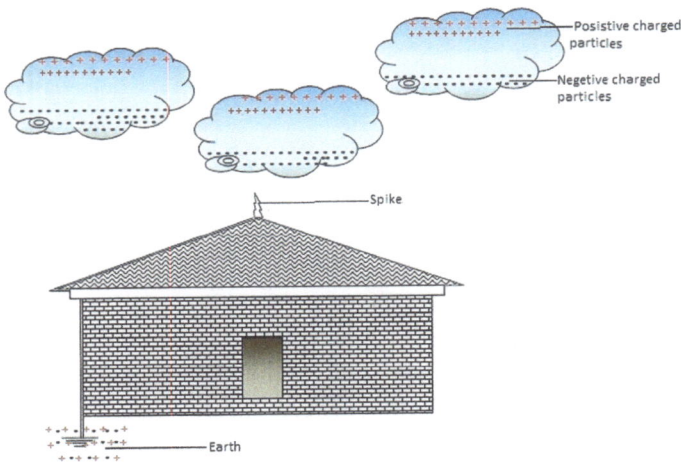

Figure: 1 The working Principle Of Thunder Protector

If a negatively charged thundercloud passes over the area of a building as shown in the diagram of Figure 1, the sharp point of the spikes gain induced charges opposite to those of the cloud from the earth.

The air around the spikes becomes ionized due to point action, which now moves upward and causes the formation of positively charge space, higher up in the atmosphere. The negatively charged electrons are attracted towards the spikes because unlike charges attract (from the basic law of magnetism) and are carried harmlessly down to the earth through the thick conducting copper cable. Thus, the house will not feel the effect of that thunderclap.

3.3   A MIRAGE

On our highways, when the atmosphere is hot, we sometimes see a pool of water, which moves further as we approach nearer to it. This pool of water is called *a mirage*. This also happens in the open Desert or a field of sand dunes

The word mirage came to English via French, mirage. From Latin *Mirare,* meaning to appear or to seem. It has the same root as mirror. Like mirror which reflects images set before it, a mirage shows images of things, which are elsewhere.

Mirage is a naturally occurring optical phenomenon in which light rays from the sun are bent to produce a displaced image of a distant object or the sky. It can be seen most frequently along an overheated highway, which disappears upon a closer viewing. It is not an optical illusion but real phenomena which is caused when light passes from colder air to warmer air, it bends away from the direction of the temperature gradient because cold air is denser then warm air and has a greater refractive index. When it passes from hotter air to cooler air, it bends towards the direction of the gradient.

If the air near the ground is warmer than that higher up, the light rays will bend upwards. Once it reaches the viewer's eye, the eye tracts it as the line of sight, which is the line tangent to the path the ray took at that point. The result is an inferior mirage from which the sky above appears on the ground. The viewer may incorrectly interpret this sight as water on the high ways.

Mirage can be classified into:

1. *Inferior*
2. *Superior*

INFERIOR in the sense that the image is seen under the real object as water or spilled oil often called desert mirage or highway mirage because it usually occurs in a sunny day on a highway or in a desert.

SUPERIOR MIRAGE occurs when the air below the line of sight is cooler than that above. In this case, the light rays are bent downwards, thus, the image appears below the true object. This can be seen when an object is kept beside a pool of water, the superior mirage will be seen below the true object in the pool of water but in an opposite direction. It is more stable than the inferior mirage.

3.4 POOR SIGNALS IN STEEL FRAME OFFICE BUILDING

The signals on our mobile phones travels in vacuum (do not require any material medium for their propagation). As such, they are electromagnetic wave.

Cellular phones in a steel framed office building sometimes have poor signal strength because since the electromagnetic wave arriving from outside the building cannot propagate or pass through a conductor (steel), but remains on its surface as postulated by Gauss in his theory, the waves are reflected outside the steel office. Thus, less wave reach the phone. As such, reception is compromised.

3.5   RADIO SIGNALS RECEPTION

In radio transmission, signals from a broadcasting station travel over great distances in vacuum with frequencies between 20Hz to 20KHz, after being subjected by one of the Modulation processes. Reception of these signals on our radio from the broadcasting station such as Rhythm Fm, Radio Nigeria and so on, undergoes series of processes that can be divided into various parts. They are:

3.5.1   TRANSMITTER:

This is housed in the Broadcasting Station; its purpose is to produce radio waves for transmission into space. The important components of this transmitter are

I.   *The Microphone*

II.   *Audio Amplifier*

III.   *Oscillator*

IV.   *The Modulator*

3.5.2   THE MICROPHONE:

This is a device that converts sound waves into electrical waves. When the signal in form of music, speeches and so on are played, the varying air pressure on the microphone generates an audio electric signal, which corresponds to the frequency of the original signal.

3.5.3   AUDIO AMPLIFIER:

The audio signal from the microphone is usually weak and requires amplification; the cascaded audio amplifier *(amplifier connected in series such that the output of the first amplifier becomes the input of the second, thus, increased the audio signals)* amplifies the output from the last audio amplifier and feeds it to the modulator for the processes of modulation.

3.5.4 OSCILLATOR:

The function of the oscillator is to produce a high frequency signal called a carrier wave; the power level of the carrier wave is raised to a sufficient level by radio frequency amplifier stages.

3.5.5 MODULATOR:
The amplified audio signal and carrier wave are fed to the modulator; the audio signal is superimposed on the carrier wave or audio waves. As the carrier frequency is very high, the audio signal can then be transmitted at large distances.

The radio waves from the transmitter are fed to the transmitting antenna or aerial from where they are radiated into space.

3.5.6 THE ANTENNA

The transmitting antenna radiates the radio waves in space in all directions. These waves travel with the speed of light (i.e. 3.0×10^8 m/s). They are electromagnetic waves and equally posses the same properties which are similar to light and heat waves except that they have longer wavelengths.

3.5.7 RADIO RECEIVER

On reaching the receiver antenna, the radio waves induce tiny electromotive force (e.m.f.) on it. This small voltage is fed to the radio receivers. The radio waves first amplified, and then the signal is extracted from them by the processes of demodulation. The signal is amplified by the audio amplifier and then fed to the speaker for conversion into sound waves. Hence, a radio receives signal of all transmitting stations at the same time but selects and interprets only the broadcast from one station with which the radio circuit is at resonance.

Fully detailed explanation of the above processes is beyond the target of this book.

3.6   THE RAINBOW

Several years ago, some people believed in the thought (superstitiously) that the rainbow is a "god". Some believed that whenever a rainbow is seen, a male lion has been born. Equally, some believed that the rainbow has at its foot a pot of gold.

In the Greek mythology, the rainbow was considered to be a path made by a messenger (Iris) between Earth and Heaven. In Chinese mythology, the rainbow has a slit in the sky sealed by goddess Nuwa using stones of five different colors.

In the Holy Bible, it is a symbol of the covenant between God and Noah; God promised Noah that he would never again flood the earth.

Scientifically, rainbows are optical and meteorological phenomena that cause a spectrum of light to appear in the sky, when the sun shines on droplets of moisture in the atmosphere. They take the form of a multi-colored arc with red on the outer part of the arc and violet on the inner section of the arc.

Its appearance is caused by dispersion of sunlight as it goes through raindrops as shown in Figure 2. The light first refracts as it enters the surface of the raindrops, reflected off the back of the droplets and again refracted as it leaves the top. The amount of light which is refracted depends upon its wavelength and thus, its color.

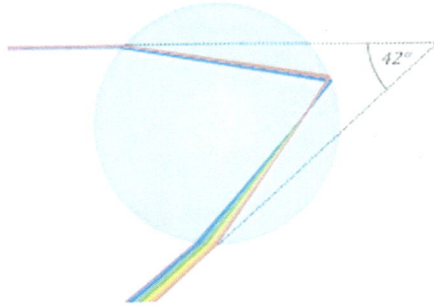

Figure 2 Dispersion of sunlight by a single raindrop

A rainbow does not actually exist at a particular location in the sky; it is instead an optical phenomenon whose apparent position depends on the observer's location and position of the sun. It has a continuous spectrum of color.

Traditionally, the full sequence of color is most commonly cited and remembered with the mnemonics "Roy G. Biv" or "Richard of York Gave Battle In Van" which means Red, Orange, Yellow, Green, Blue, Indigo and Violet.

CHAPTER FOUR

PHYSICS AND SOME OTHER RELATED SCIENCES

The science of Physics has developed out of the efforts of men and women to describe how and why our physical environment behaves, These efforts have been so successful that today, Physics encompasses remarkably diverse phenomena. It relates to the planets orbiting the sun, transistors working in radios and computer systems, laser beam used in Industries and Surgery and so on

The key to understanding the strength of Physics is to recognize that their laws are based on experimental facts. Its exciting feature is its capacity of predicting how nature behaves in one situation, on the basis of experimental data obtained in another situation. Such predictions place Physics at the heart of modern Technology and adds great value to our lives.

Rocketry and the development of space travel have their roots firmly planted on the physical laws of **Galileo Galilee and Sir Isaac Newton**. The transport industry relies heavily on the principles in Physics for the development of engines and design of thermodynamic vehicles. All electronic and computer industries owe their existences to the impetus provided by the inventions of transistors and integrated circuits (ICs), which grew directly from physical laws describing the electrical behavior of solids.

The Telecommunication industry depends extensively on electromagnetic waves whose existences were predicted by **James Clerk Maxwell** in his theory of Electricity and Magnetism. The medical profession uses X-rays, ultrasonic and magnetic resources methods for obtaining images of the internal parts of the human body, and Physics lies at the core of all these. Above all, Physics has relationships with all other sciences such as Mathematics, Geology, Chemistry, Biology, Medicine, and Engineering.

4.1 PHYSICS AND MATHEMATICS

The laws of Physics are usually expressed as mathematical equations, which are then used to make predictions about other phenomena. For instance, the Newton's Law of gravitation, which states that:

> "*The force of attraction between two objects is proportional to the masses of each other and inversely proportional to the square of the distance between them*"

Mathematically, it can be restated as $F = \dfrac{Gm_1m_2}{R^2}$

Where F = Force
G = Gravitational Constant (6.67×10^{-11} Nm^{-2})
m_1 m_2 = Masses of the Objects
R = Distance between the Objects

Equally, the second law of motion, which states, "*If an external force acts on a body, the body accelerates, the direction of acceleration is the same as the direction of the applied force*".

The net force vector is equal to the mass of the body multiplied by the acceleration of the body. Mathematically, it can equally be expressed as.
F = MA
Where F= Force acting on the body
M = Mass of the body
A = Acceleration of the body due to the force acting on it.

This laws of gravity together with other laws in Physics like the laws of motion can be used to predict the orbit of Planets, Comets and

Satellites. Understanding such predictions require knowledge of elementary calculus and the ability to solve simple and partial differential equations.

It is usually easier to study Physics and Mathematics at the same time, since the immediate application of mathematics to physical situation helps to understand both physics and mathematics. Equally mathematics deals with abstract ideals but physics brings those ideals to realities through experiments. Hence, ***Mathematics is the language of physics.***

4.2   PHYSICS AND GEOLOGY

Geologists and Physicists (Geophysicists) have made many profound discoveries over the past decades about how the earth functions; including the existence of Tectonic Plates, the dynamics of Volcanoes and Earthquakes, the general structure of the earth's deep interior and details of its surface layers.

In Oil and Gas sector, oil and gas are trapped on porous rocks that have low densities. Knowing this fact enables the scientist (geophysicist and geologist) to search for such an oil bearing strata. They explore the interior of the earth's surface layer or crust in a variety of ways such as seismic, gravity, magnetic, electrical resistance and induced polarization methods.

In seismic methods, setting off a series of explosion on the surface of the earth or ocean bottoms, and then detecting the vibration of the earth caused by the explosion do seismic reflecting proofing. The energy of the blast is transferred into the earth as longitudinal waves (like sound wave) and transverse waves, traveling down a string. These waves travel through various layers of the earth reflecting and refracting as they encounter rocks of different density. Some of these reflected and refracted waves are picked up by sensitive microphone called **geophones** or **hydrophones** (in case of an ocean), which are usually placed around the region where the explosion occurred.

Timing the returns of the waves from different explosion sites, and knowing the speed of the wave enables the scientists to determine the interior distribution of rock in the earth's crust which equally helps to determine where to look for the Oil.

The crucial point to remember is that the principle behind these waves reflecting and refracting lies on basic physics, which is beyond the target of this book.

4.3   PHYSICS IN MEDICINE

The knowledge of Physics in Medicine has made great impacts in the present day, than it was in the medieval age (olden days). Consider the physics that is involved in a medical emergency. Supposing there was a serious accident in which one received a head and abdominal wound, one is rushed to the hospital in an ambulance powered by an internal combustion engine, one's wounds are immediately x-rayed and the film developed within a minute. The doctor examines the negative film through a screen that is back – lit by fluorescence light; meanwhile, one is connected to a Cathode Ray Tube (CRT) monitor to detect the changes in internal body.

The fundamental Physics discovered over the century played a crucial role in the engine of the ambulance, using the physics of combustion and thermodynamics. The air in the emergency room was electrostatically cleaned to prevent infection; the x-ray machine is as a result of the fundamental research in Physics. The CRT monitor is a simplified television monitor, which is based on the atomic theory of element, Ohm's law of Electrical Circuit and Electromagnetism.

Another important application of the knowledge of Physics in Medicine is the series of internal imaging devices such as

I.	Computerized Axial Tomogarphy (CAT)
II.	Magnetic Resonance Imaging (MRI)
III.	Positron Emission Topography (PET) and so on

These apply the theories of electromagnetism, atomic and nuclear physics and computer technology to allow physicians see the details of the inner working of the human body.

The advances in our understanding of Physics have enabled medical researchers to understand many of the functions of the human body. As a result of this, medicine now provides us with cure for many of the most deadly diseases that have claimed thousands or even millions of lives, in the past

If these advances in Physics made this past century had not come, most of these devices which are being used in medicine would not have existed. Thus, medicine would still be a medieval art and the quality of lives would be much lower.

4.4   PHYSICS IN CRIMINOLOGY

Criminology is the science that studies crime in a given society. Thousands of years ago when people started living together in large communities, they realized that they needed protection not only from outside their neighborhood but also from members of their own population. As a result of this, the police force came to existence.

At the end of the twentieth century, with higher population densities than before, and an increase in crime rate, the need for effective protection became the order of the day. Fortunately the application of physics to criminology has strengthened law enforcements.

For instance, for one to enter a Bank for a transaction, one has to pass through a scanning machine (metal detector), which uses principles in Physics to detect metal or gunpowder, such that robbers or criminals would not enter the Banking Hall with arms.

In politics, a device has now been developed as a means of checking crimes during Elections by using Electronic Voting machines which are just computers that use semi-conductors, transistors assembled to detect individual's prints (finger-print).

Our "high tech" society and all its electrical and mechanical devices, relies on the enormous foundation of the basic principles in Physics to function and grow.

CHAPTER FIVE

CAREERS IN PHYSICS AND EMPLOYMENT OPPORTUNITIES

Most often, students pursue careers that are not suitable to them. Sometimes, parents force their children into the University to accomplish a dream they could not achieve in their youthful age, due to one reason or the other. When these students eventually gain admission to read these Courses, they might not find them easy.

There is one incidence that happened in a place; a science student was instructed by his parents to change and read law in one of our citadels of learning (University) because they already have a medical doctor, an engineer and hope to get a lawyer when this student graduates. This boy accepted the decision and enrolled for the course (i.e. law). During examinations as a law student, he could not write extensively, to show the lecturer that he knew the question, he usually summarized everything as a science student. Before this boy could know what was happening, his Head of the Department (HOD) had (already) listed his name as one of those they were encouraging to leave the department because he could not handle the number of "carry overs" he got within two semesters. Finally, this boy left that university to write the UME examination in order to read the Course he was more suitable for.

There are thousands of things to discover, the reason you are created is to do what has not been done. There is a vacuum that needs to be filled by you alone, if everybody wants to be a doctor, who will then produces what the doctors will eat. No discipline is more important than the other, as no part of the human body is more important than the other.

The question is how we discover these God-given potentials that are latent in us? How do we know when we are on the right track? There are certain questions one needs to ask oneself before choosing a career. For instance, if one want to be a scientist, one need to ask oneself the following questions.

5.11   ARE STUDIES IN PHYSICS FOR ME?

If you enjoy learning and want to really understand things around you, you like mathematics and experiments, then you should consider becoming a physicist (scientist); you will find the subject fascinating, the ideas themselves will motivate you to study and learning will become an enjoyable task.

5.12   WHAT TRANING WILL IT PROVIDE?

The most famous scientists such as Albert Einstein, Galileo Galilee, and Isaac Newton were physicists. Physicists are one of the most highly trained scientists, versed in mathematics and design of complex instruments. As such, they are also one of the most versatile scientists able to easily cross boundaries into other disciplines such as chemistry, biology, medicine, etc. There are biophysicists, geophysicists, medical physicists etc. When one has a deep understanding of nature, it is easy to apply that understanding in a variety of ways.

5.13   WHAT CAREER OPTIONS WILL PHYSICIST HAVE?

Physicists end up in all sort of interesting jobs and are virtually never unemployed because of their broad training and adaptability. Particularly, they are needed at the start of new technologies and in challenging projects. They are needed in oil and gas industry as geophysicist to analysis the reflection and refraction of waves during oil exploration (using Seismic methods).

There is no field in Industry that does not or cannot employ knowledge obtained from Physics in one way or the other, but the progress of this will depend on the academic and personal qualities of the physicist.

Below are some industries where physicists and knowledge of Physics are employed:-

INDUSTRY	RELATED FIELD
Telecommunication Industry	Microwave Physics
Power and Energy	Electricity, Solar and Nuclear Physics
Airways and Air Force Mechanics	Astrophysics, Space and Geophysics
Engineering Industry	Instrumentation, Electronic and Mechanics
Computer Industry	Solid-State Physics and Industrial Electronics
Oil Industry and Solid Minerals	Geophysics and Earth Physics
Steel Industry	Solid State and Mineral Science
Metrological Industry	Physics of Atmosphere
Gas Company	As "in Matter Technologist"
Research Institute	As a Scientist
Construction Firm	As in Mechanic and Physics of Solid Minerals
Ministry of Education	As a Trainee Physicist

Ministry of Science and Technology	As a Core Scientist and Technologist
Maintenance and Management Sector	As a Quality Control Technologist.

From the above, it will be discovered that a lot of job opportunities await a physicist on graduation.

As a working physicist, you may find yourself trying to predict the stock market, testing satellites, developing new electronic devices and components, doing medical physics in a hospital, teaching the next generation of physicists in High Schools or even predicting the next eclipse that will occur in Nigeria, Africa or the world at large.

CHAPTER SIX

PHYSICS IN DIFFERENT CAREERS

There is scarcely an area of modern life or career where the knowledge gained from studying Physics cannot be applied. Careers such as Journalism, Athletics, Auto-Mechanic, Imaging technician and so on, apply some of these Principles, Theories and Laws propounded by our past scientists in order to do their works easily and more effectively in the following ways:

6.1   JOURNALISM

The knowledge of science (Physics) is one of the best assets a reporter can have. New discoveries and findings, space researches and politics, medical breakthroughs, natural disaster, technological competitiveness, interviews, the environment, and so on. make up the larger part of the News. Reporters who have a basic background in Physics have an advantage in being able to grasp technical issues quickly and communicate easily with Researchers.

Many major daily newspapers in our country have Science Sections.

6.2   ATHLETICS

Whenever I watch an athlete, I see the Principles of Physics in motion. A basketball player for instance, knowing the angle at which the net is to his position with the ball can help him throw the ball into the net with more accuracy, because the path taken by the ball when thrown (trajectory) is a phenomenon in Physics (projectile motion).

The bat-hitting of the base ball, the spiraling football, the bend in the vaulter's pole and the tension of muscles as a weight is lifted illustrate some of the basic laws in Physics like momentum, equilibrium, velocity, kinetic energy, center of gravity, projectile motion and friction. Knowing these principles helps the athlete to improve his performances.

6.3 IMAGING TECHNICIAN

Imaging technicians work at hospitals, medical colleges and clinics. Looking inside the body without surgery is one of the most important tools an image technician uses. He uses X-ray, computed topography (CT) scans and magnetic resonance imaging (MRI) to determine the shape of broken bones, diagnose diseases and develop treatment of various illnesses such as cancer and so on.

Technicians who use imaging equipment need to be familiar with the concepts of X-rays, photoelectric Effects, Compton Effects and Magnetic Resonance and be able to determine how these technologies are used.

6.4 AUTO-MECHANICS

A look on our road today have shown that automobiles are far away from those put on the road by *Henry Ford,* computers play an important role on how our automobiles operate. These computers are just integrated semi-conductors such as diodes, rectifiers, capacitors, resistors, etc. embedded to do certain work such as igniting, opening of doors, locking, etc an automobile.

Knowing the principle behind these semi-conductors and how they work is essential to an auto mechanic.

CHAPTER SEVEN

SOME CHRONOLOGICAL DISCOVERIES IN PHYSICS

1881	Michelson obtains null result for absolute velocity of the Earth.
1884	Balmer finds empirical relationship for spectral lines of Hydrogen
1887	Hertz produced electromagnetic waves, verifying Maxwell's Theory and accidentally discovering Photoelectric Effects
1887	Michelson repeated his experiment with Mosley again obtaining null result
1895	Roentgen discovered X-ray
1896	Berquerrel discovered Nuclear Radioactivity
1897	J.J. Thomson measured charges to mass ratio for cathode rays showing that electrons are the fundamental constitutions of atoms
1900	Planck explains Blackbody Radiation using energy quantization involving new constant (h)
1900	Lenard investigated photoelectric effects and fines energy of electron independent of light intensity.
1905	Einstein proposed Special Theory of Relativity and explained the photoelectric effect by suggestion of quantization of radiation.
1907	Stein applied energy quantization to explain temperature dependence of heat capacity of solids.
1908	Rydberg and Ritz generalized Balmer's formula to fit spectral of many elements.
1909	Millikan's oil drop experiments show quantization of electric charge.
1911	Rutherford proposed nuclear model of atom based on Alpha particles for Scattering experiment of Geiger and Marsden.

1912	Friedrich Knipping and Von Lane demonstrated diffraction of X-rays by crystal showing that x-rays are waves and crystals are regular arrays.
1913	Bohr proposed model of Hydrogen Atom
1914	Mosley analyses X-ray spectral, using Bohr's model to explain Periodic Table in terms of Atomic Numbers.
1914	Frank and Hertz demonstrated atomic energy quantization.
1915	Duane and Hunt showed that the short wavelength unit of X-rays is determined from quantum theory.
1916	Wilson and Summerfield proposed rules of quantization of periodic system.
1916	Millikan verified Einstein Photoelectric Equation.
1923	Compton explained X-ray Scattering by electrons as collision of photon and electron and verified results experimentally.
1924	De-Broglie proposed Electron Waves of the wavelength.
1925	Edwin Schrödinger developed mathematics of Electron Wave Mechanics.
1927	Heisenberg formulated Uncertainty Principles.
1927	Davison and Germer observed electron waves Diffraction by Single Crystal.
1927	G.P. Thomson observed Electron Wave Diffraction in metal foil
1928	Gamow and Condon applied quantum mechanics to explain Alpha-Decay Lifetime.
1928	Dirac developed Relativistic Quantum Mechanics and predicted the existence of Positron.
1932	Chadwick discovered Neutron
1932	Anderson discovered Positron.

REFERENCES

1. Dull, C.E. and William J.E. (1960) *Modern Physics,* Cambridge University Press.
2. Smoot, R.C. and Murphy J.T. (1982) *Principle of Physics and Problem*, Dover Publication.
3. Bransden, B.H. (2002) *Physics of Atom and Molecules* Prentice Hall.
4. Lee Raymond, L and Alistair B.P (1980) *The Rainbow Bridge*, Pennsylvania State University Press.
5. Ugwuoke, A.C. (2005) *Modern Optics*, P and C publisher Enterprises.
6. Mehta, V.K. and Mehta R (1980) *Principle of Electronics*, S. A. Chard and Company LTD.
7. Anyakoha M.W. and Okeke P.N. (1987) *Physics for Senior Secondary Schools* Macmillan Education LTD.
8. Young H.D. and Freedmen R.A. (2004) *University Physics* Pearson Education (Singapore) LTD.
9. Stem P.D. (1956) *Our Space and Environment*, HOH, Rinehart and Winston Inc. New York.
10. Minnaret, M.P (1973) *The nature of Light and Color in the Open Air* Dover Publication.
11. David L.K. and William L.P. (2001) *Color of Light in Nature* Cambridge University Press.
12. Griffith, David .J. (1998) *Introduction to Electrodynamics* 3rd Edition. Prentice Hall.
13. McQuarries, Donald (2000) S*tatistical Mechanics* University Science Book.
14. Chandler, David (1987) *Introduction to Modern Statistical Mechanics,* Oxford University Press.
15. Paul .G. (1993) *Conceptual Physics*, Donnelley and Sons Company
16. Clark .H. (1974) *First Course in Quantum Mechanics,* Van Nostrand Reinhold Company LTD.

www.ingramcontent.com/pod-product-compliance
Lightning Source LLC
Chambersburg PA
CBHW041718200326
41520CB00001B/145